ISBN 978-1-334-41129-8
PIBN 10739314

This book is a reproduction of an important historical work. Forgotten Books uses
state-of-the-art technology to digitally reconstruct the work, preserving the original format
whilst repairing imperfections present in the aged copy. In rare cases, an imperfection in
the original, such as a blemish or missing page, may be replicated in our edition. We do,
however, repair the vast majority of imperfections successfully; any imperfections that
remain are intentionally left to preserve the state of such historical works.

1 MONTH OF
FREE
READING

at

www.ForgottenBooks.com

By purchasing this book you are eligible for one month membership to ForgottenBooks.com, giving you unlimited access to our entire collection of over 1,000,000 titles via our web site and mobile apps.

To claim your free month visit: www.forgottenbooks.com/free739314

STUDY OF THE DISTRIBUTION OF IODINE BETWEEN CELLS AND COL- LOID IN THE THYROID GLAND

•

A DISSERTATION

SUBMITTED TO THE FACULTY
OF THE OGDEN GRADUATE SCHOOL OF SCIENCE
IN CANDIDACY FOR THE DEGREE OF
DOCTOR OF PHILOSOPHY

DEPARTMENT OF PHYSIOLOGICAL CHEMISTRY AND PHARMACOLOGY

BY

HARRY BENJAMIN VAN DYKE

Private Edition, Distributed By
THE UNIVERSITY OF CHICAGO LIBRARIES
CHICAGO, ILLINOIS

Reprinted from
he JOURNAL OF BIOLOGICAL CHEMISTRY, Vol. XLV, No. 2, 1921; Vol. LIV, No. 1, 1922
The AMERICAN JOURNAL OF PHYSIOLOGY, Vol. LVI, No. 1, 1921

A STUDY OF THE DISTRIBUTION OF IODINE BETWEEN CELLS AND COLLOID IN THE THYROID GLAND

A DISSERTATION

SUBMITTED TO THE FACULTY
OF THE OGDEN GRADUATE SCHOOL OF SCIENCE
IN CANDIDACY FOR THE DEGREE OF
DOCTOR OF PHILOSOPHY

DEPARTMENT OF PHYSIOLOGICAL CHEMISTRY AND PHARMACOLOGY

BY

HARRY BENJAMIN VAN DYKE

Private Edition, Distributed By
THE UNIVERSITY OF CHICAGO LIBRARIES
CHICAGO, ILLINOIS

Reprinted from
The JOURNAL OF BIOLOGICAL CHEMISTRY, Vol. XLV, No. 2, 1921; Vol. LIV, No. 1, 1922
The AMERICAN JOURNAL OF PHYSIOLOGY, Vol. LVI, No. 1, 1921

Reprinted from The Journal of Biological Chemistry, Vol. XLV, No. 2, 1921

A STUDY OF THE DISTRIBUTION OF IODINE BETWEEN CELLS AND COLLOID IN THE THYROID GLAND.

II. RESULTS OF STUDY OF DOG AND HUMAN THYROID GLANDS.

By HARRY BENJAMIN VAN DYKE.

(From the Laboratory of Physiological Chemistry and Pharmacology, University of Chicago, Chicago.)

(Received for publication, November 29, 1920.)

In the first paper of this series Tatum[1] described a method whereby thyroid cells may be separated from colloid material and examined chemically. Briefly the method consists in cutting frozen sections of the thyroid gland and floating these sections on Ringer's[2] solution. The colloid material immediately drops out of the acini and is presumably dissolved in the Ringer's solution. The cells may then be separated by centrifugalization, dried, weighed, and analyzed. Comparisons between the iodine content of cells so separated and the iodine content of control pieces of unsectioned whole gland may indicate the distribution of iodine between cells and colloid under different functional conditions.

In this paper I wish to report the results of a study, suggested by Dr. A. L. Tatum, of the distribution of iodine in the thyroid glands of normal and iodine-fed dogs as well as in human glands obtained from individuals subjected to operation for toxic goiter. The method earlier described by Kendall[3] was used in making the final iodine determinations.

Incidental to the determination of the iodine distribution in dog and human thyroid glands, some control experiments were performed relative to the alterability of the intracellular iodine concentration during the process of cutting. It may be argued that a portion of the iodine-containing compound diffuses from

[1] Tatum, A. L., *J. Biol. Chem.*, 1920, xlii, 47.
[2] Campbell, J. A., *Quart. J. Exp. Physiol.*, 1911, iv, 1, Formula "A."
[3] Kendall, E. C., *J. Biol. Chem.*, 1914, xix, 251.

325

cells as they lie suspended in Ringer's solution during the separation of the colloid material. If it is assumed that such an outward diffusion takes place, it seems reasonable to expect that the iodine compound should diffuse back into the cells if we should increase the concentration of that iodine compound in the Ringer's solution in which the cells are suspended. Several experiments like those given in Table I were undertaken to determine whether or not the iodine content of the cells could thus be increased.

TABLE I.

Effect of Floating Cells on Pure Ringer's Solution and on Ringer's Solution on Which Cells of an Iodine-Rich Gland Had Been Floated.

Animal No.	Weight of whole gland used.	Iodine in whole gland.	Weight of cell mass used.	Iodine in cell mass.	Ratio of percentage of iodine in cells to percentage of iodine in whole gland.	Remarks.
	mg.	*per cent*	*mg.*	*per cent*		
14	146.0	0.187	40.5	0.017	0.091	Cells floated on pure Ringer's solution.
			53.0	0.017	0.091	Cells floated on Ringer's solution on which previously cells of iodine-rich gland had been floated.
20	287.2	0.030	161.5	0.005	0.167	Cells floated on pure Ringer's solution.
			151.0	0.006	0.20	Cells floated on Ringer's solution on which previously cells of iodine-rich gland had been floated.

It may be seen from the data given in Table I that no increase in the iodine content of cells could be brought about by increasing the concentration of the characteristic iodine compound in the Ringer's solution.

Again the iodine concentration of the cell mass does not seem to be in the least altered whether or not the freshly centrifugalized cell mass is washed several times with iodine-free Ringer's solution.

Moreover, in the attempt to find a suspending medium more nearly related physicochemically with thyroid cells than is Ringer's solution, I have used fresh dog serum and have not found the iodine content of the cells measurably different from that of control cells suspended in Ringer's solution.

Finally autolysis does not seem to be much of a factor in the loss—if any occurs—of iodine compounds from cells as they lie in contact with Ringer's solution. Throughout the process of cutting and separating the cells, the reaction of the suspending medium should remain weakly alkaline—a reaction which has been shown to be unfavorable to autolysis.[4] Also in experiments to be reported later in which the same technique was used under slightly different·conditions, every effort was made to eliminate autolysis by cutting successively small portions of the gland and floating the cells on Ringer's solution cooled by ice. The cells from the small portions cut successively were at once centrifugalized and dried. No change could be noticed in the iodine distribution.

Results of Study of Dog Glands Taken at Random.

TABLE II.

Quantitative Determination of Iodine in Whole Gland and in Cells Free from Colloid Material of Thyroid Glands of Normal Dogs.

Animal No.	Weight of whole gland used.	Iodine in whole gland.	Weight of cell mass used.	Iodine in cell mass.	Ratio of Percentage of iodine in cells to percentage of iodine in whole gland.	Morphology.	
						Cells.	Colloid.
	mg.	*per cent*	*mg.*	*per cent*			
5	550.0	0.25	154.0	0.031	0.124	Flat.	Fair.
6	388.0	0.175	265.0	0.027	0.154	"	Rich.
10	480.0	0.017	342.0	0.003	0.176	Cuboidal.	Poor.
11	505.5	0.032	197.5	0.005	0.156	"	"
12*	278.7	0.006	286.5				
13	689.0	0.234	154.0	0.047	0.201	Flat.	Fair.
14	146.0	0.187	40.5	0.017	0.091	"	"
19	474.5	0.024	116.0			"	Poor.
20*	287.2	0.030	161.5	0.005	0.167		
21*	604.7	0.038	189.0	0.006	0.158		
23	458.2	0.003	390.4			Cuboidal.	Poor.
24	318.7	0.021	295.4			"	"
31	513.8	0.055	205.6	0.008	0.145	"	"
32*	235.2	0.179	103.9	0.037	0.207		
33	289.3	0.011	538.3	0.002	0.182	Flat.	Poor.

* No histological examination.

[4] Bradley, H. C., J. *Biol. Chem.*, 1915, xxii, 113. Bradley, H. C., and Taylor, J., J. *Biol. Chem.*, 1916, xxv, 261.

From the results given in Table II one sees that the ratio of iodine concentration in cells to iodine concentration in whole gland (and hence the ratio of iodine concentration in cells to colloid-iodine concentration) has a quite constant value. In some glands (Nos. 12, 19, 23, and 24), to be sure, the iodine content of the cell mass analyzed was so low that no ratio value could be obtained. But in general, despite great variations in the iodine content and morphology of the glands analyzed, the ratio values change relatively little. These findings in the dog's thyroid gland are similar to those of Tatum[1] in the beef, pig, and sheep thyroid glands. A comparison of the ratio values in different animals is given below.

Effect of Feeding Iodine.

It was thought that perhaps ratio changes could be induced by the administration of iodine or iodine compounds. Capsules containing 1 gm. of potassium iodide or 2 drops of tincture of iodine in starch were fed over varying periods of time. No more than one capsule was given in 24 hours. The results of the examination of the glands of animals so treated are given in Table III.

The ratio obtained from analyzing the cells and whole gland of the thyroid glands of animals which had received varying amounts of iodine or iodine compounds over periods of no less than 3 days is practically the same as that of the animals which had received no iodine. One gland (that of No. 16) gave an unusually high value which I cannot explain. In view of the variations in morphology and iodine content of the glands of the normal series together with the quite constant ratio exhibited by that series, it is not surprising that the ratio still remains constant despite the feeding and consequent absorption of iodine. From a consideration of the glands examined it appears that the ratio is not altered by the feeding of iodine as potassium iodide or free iodine over periods of time ranging from 3 days to 3 weeks; and yet the total iodine content of these glands undoubtedly is greatly increased during that same period.

TABLE III.

Quantitative Determination of Iodine in Whole Gland and in Cells Free from Colloid Material of Thyroid Glands of Iodine-Fed Dogs.

†Animal No.	Form of iodine.	Period of feeding.			Weight of whole gland used.	Iodine in whole gland.	Weight of cell mass used.	Iodine in cell mass.	Ratio of percentage of iodine in cells to percentage of iodine in whole gland.
		Daily.	On alternate days.*	Total.					
		days	*days*	*days*	*mg.*	*per cent*	*mg.*	*per cent*	
2	KI	14	0	14	98.2	0.31	77.7	0.045	0.145
7	KI	14	7	21	147.0	0.12	113.0	0.031	0.258
8	KI	14	7	21	124.5	0.87	125.0	0.080	0.092
9	KI	14	2	16	353.5	0.251	116.0	0.040	0.159
16	KI	12	0	12	651.4	0.319	316.1	0.117	0.367
17	KI	12	0	24	646.5	0.617	302.3	0.083	0.134
	Tincture of I.	12	0						
27	" " I.	3	0	3	556.1	0.302	332.8	0.046	0.152
29	" " I.	3	0	3	195.2	0.314	158.6	0.039	0.124
30	" " I.	3	0	3	163.3	0.357	127.5	0.041	0.115

* After feeding daily.

Human Glands.

The distribution ratio of iodine between cells and colloid material was determined in thirteen human glands obtained from operative cases.[5] In Table IV are given the results of the analyses of the human glands together with the clinical diagnosis made in connection with ten of the cases.

It will be seen that most of the thyroid glands reported were clinically diagnosed as toxic goiters. Considerable variations in the ratio value occur and do not appear to be related either to the iodine content of the gland or to its morphology. The ratio value of No. 22 is inexplicably high. In the human gland series, however, as in other gland series previously reported, the ratio variations are of a much smaller magnitude than the variations in total iodine content. So here too the ratio is fairly constant despite variations in morphology and iodine content.

[5] Through the courtesy of Dr. C. B. Davis and Dr. A. D. Bevan of the Presbyterian Hospital, Chicago, and of Dr. A. J. Ochsner of Augustana Hospital, Chicago.

TABLE IV.

Quantitative Determination of Iodine in Whole Gland and in Cells Free from Colloid Material of Human Thyroid Glands.

Series No.	Weight of whole gland used.	Iodine in whole gland.	Weight of cell mass used.	Iodine in cell mass.	Ratio of percentage of iodine in cells to percentage of iodine in whole gland.	Morphology.		Diagnosis.
						Cells.	Colloid.	
	mg.	per cent	mg.	per cent				
22	284.7	0.081	260.8	0.049	0.605	Cuboidal.	Poor.	Colloid cystic goiter with toxic symptoms.
25*	600.2	0.274	308.6	0.105	0.383	Flat.	Rich.	Mild exophthalmic goiter.
	411.0	0.276	240.7	0.098	0.355			
35	671.8	0.286	179.2	0.051	0.178	Flat.	Rich.	Toxic thyroid.
36	439.7	0.158	219.9	0.020	0.127	Cuboidal.	Fair.	" goiter.
37†	667.5	0.092	258.4	0.027	0.293			Colloid-goiter.
38	578.6	0.079	191.7	Trace.		Cuboidal.	Poor.	" " following parenchymatous hyperplasia of exophthalmic goiter.
39†	516.0	0.279	297.6	0.095	0.341			
40	469.2	0.345	225.5	0.066	0.191	Flat.	Rich.	Exophthalmic goiter.
41	559.2	0.280	187.0	0.067	0.239	"	"	
48*	549.9	0.043	312.4	0.005	0.116	Cuboidal.	Poor.	
	607.3	0.044	311.3	0.005	0.114			
49*	590.7	0.137	256.0	0.026	0.189	Cuboidal.	Fair.	Exophthalmic goiter.
			204.5	0.027	0.197			
51	731.4	0.283	294.4	0.072	0.254	Flat.	Rich.	" "
60	370.0	0.152	228.0	0.015	0.099	"	"	Cystic goiter with toxic symptoms.

* Duplicate determinations made.
† No histological examination.

Comparison of Ratio Value in Different Animals.

A comparison of the ratio values of the thyroid glands of the different animals so far examined is given in Table V. The few abnormally high and unexplained ratio values are not included in the table.

From what data are available there appear to be some differences in the thyroid glands of different animals in the numerical value of the ratio of the percentage of iodine in cells to that in whole gland. The ratio value of iodine distribution for dog thyroid glands seems to be consistently lower and more constant than that for the thyroid glands of the other animals studied.

TABLE V.

A Comparison of the Value of the Ratio of the Percentage of Iodine in Cells Free from Colloid Material to the Percentage of Iodine in Whole Gland in Different Animals.

Animal.	Extremes of iodine content of whole gland.	Ratio values.		
		Extremes.	Mean.	Average.
	per cent			
Beef*	0.023–0.468	0.21 –0.48	0.35	0.36
Dog	0.011–0.870	0.091–0.258	0.175	0.154
Man	0.043–0.345	0.099–0.384	0.242	0.22
Pig*†	0.377–0.810	0.20 –0.34	0.27	0.27
Sheep*	0.089–0.442	0.23 –0.41	0.32	0.33

* Tatum.[1]

† Only two glands analyzed.

SUMMARY.

1. The method described by Tatum[1] was used to determine the ratio of the percentage of iodine in cells to the percentage of iodine in whole gland in the thyroid glands of normal and iodine-fed dogs as well as in human thyroid glands obtained from operative cases.

2. Evidence is presented indicating that the concentration of intracellular iodine is independent of the suspending medium, whether that is pure Ringer's solution, Ringer's solution containing iodine-rich colloid material, or homologous blood serum.

3. The ra
great variat
glands exa
seems quite
sheep.

Reprinted from THE AMERICAN JOURNAL OF PHYSIOLOGY
Vol. 56, No. 1, May, 1921

A STUDY OF THE DISTRIBUTION OF IODINE BETWEEN CELLS AND COLLOID IN THE THYROID GLAND

III. THE EFFECT OF STIMULATION OF THE VAGO-SYMPATHETIC NERVE ON THE DISTRIBUTION AND CONCENTRATION OF IODINE IN THE DOG'S THYROID GLAND

HARRY BENJAMIN VAN DYKE

From the Laboratory of Physiological Chemistry and Pharmacology, University of Chicago

Received for publication February 5, 1921

For many years it has been held that the thyroid gland is supplied with true secretory nerves. In support of this assertion there is considerable anatomical evidence and some physiological evidence. Of late the nerves which anatomists have traced into the thyroid gland and have considered to be possibly secretory in function have been declared to be branches of the cervical sympathetic nerve. And recent physiological work has tended to confirm this view. Only twelve years ago Wiener (1) published the report of experiments from which he concluded that extirpation of the inferior cervical ganglion produces a marked atrophy of the thyroid gland on the side of the extirpation. Wiener maintained that no comparable effect on the lobe of the thyroid gland on the side of the operation could be produced by vagotomy or by removal of the superior cervical ganglion. More recently Rahe et al. (2) announced that they were able to produce a quite marked diminution in the iodine concentration of the lobe of the thyroid gland on one side by stimulating the thyroid nerves in several different ways. They stimulated the nerves of the superior thyroid artery, the intact vago-sympathetic nerve as well as the vago-sympathetic nerve near the level of the superior cervical ganglion after ligating the nerve low in the neck and cutting the nerve central to the point of stimulation. They found that the most marked loss was brought about by the stimulation of the intact vago-sympathetic nerve.

Watts (3) undertook to find out whether or not the results obtained by Rahe, Rogers, Fawcett and Beebe might be due to vasomotor changes in the gland on the stimulated side. Watts likewise found that he could reduce the iodine content of the right or left lobe of the thyroid gland of the dog by stimulating the "cervical sympathetic isolated from the vagus sheath" and the "nerve filaments accompanying the superior thyroid vessels." However Watts maintained that he could cause some diminution in the iodine content simply by periodically reducing the blood flow through the gland by "occluding the main thyroid artery" the nerves of which had been dissected away. Hence he concluded that all of the effects of stimulation on the iodine content can be accounted for by the coincident vasomotor changes which he showed to be present.

Positive evidence of the secretory effect of sympathetic stimulation has been reported by Cannon and his co-workers (4) in several communications. Working with cats they sutured the phrenic nerve with the cervical sympathetic nerve and observed following the operation increased basal metabolism, respiratory hippus and falling hair which they interpreted as the results of hypersecretion of the thyroid gland caused by the periodic bombardment of the gland by impulses carried from the respiratory center to the gland's secretory nerves. Cannon and Cattell (5) adduced additional evidence as to the rôle of the sympathetic nerves in experiments dealing with the electrical condition of the gland. Following the stimulation of the upper thoracic sympathetic nerves or the injection of epinephrin they were able to show a definite action current in the thyroid gland after a latent period of five to seven seconds. Recently Cannon and Smith (6) maintained that gentle massage of the thyroid gland or stimulation of the cervical sympathetic nerve increases the rate of the denervated heart. The denervated heart is said not to be affected when the cervical sympathetic nerve is stimulated after removal of the thyroid gland. Moreover Levy (7) observed that the pressor effect of epinephrin after a variable latent period is increased by the stimulation of the cervical sympathetic nerve. He declared that stimulation of the cervical sympathetic nerve has no such effect after thyroidectomy.

However the conclusions based on the experiments mentioned above have not been universally accepted. Burget (8) was unable to alter the thyroid gland noticeably either by uniting the phrenic and cervical sympathetic nerves or by removing a section of the cervical sympathetic nerve. Marine, Rogoff and Stewart (9) sutured together the phrenic and cervical sympathetic nerves in several cats. They demonstrated a functional union between the phrenic and cervical sympathetic nerves but observed no exophthalmos, tachycardia or respiratory hippus in their animals. There was no apparent difference either grossly or histologically in the lobes of the thyroid gland on the operated and the non-operated sides. Troell (10) reported that he was unable to produce either exophthalmos or respiratory hippus by suturing the proximal end of the phrenic nerve to the cervical sympathetic nerve. Employing cocaine as a sensitizer for sympathetic nerve endings, Mills (11) did not observe, following the repeated injection of cocaine, any alteration in the amount or nature of the thyroid secretion as judged by what is known of the gland's histology. Finally Rogoff (12) records one experiment in which he drew blood from a vein of the left lobe of the thyroid gland and at the same time stimulated the cervical sympathetic nerve on that side in the hope of increasing the secretory activity of the stimulated lobe. From the right lobe he also collected blood by way of a vein. While drawing the blood he massaged the right lobe to some extent but did not stimulate the right cervical sympathetic nerve. He found that specimens of dried blood from each lobe were potent when fed to tadpoles. But he could detect iodine chemically only in the blood from the non-stimulated lobe. Moreover the non-stimulated lobe had a lower iodine content than the stimulated lobe.

In connection with some studies on the distribution of iodine in cells and colloid in the thyroid gland I attempted to alter acutely the total iodine content of the gland by stimulating the vago-sympathetic nerve of the dog. Some inconsistencies in the results in the early part of the work forced me to investigate the matter more carefully and to repeat the work of Rahe et al. (2) and of Watts (3).

Methods. Dogs were used in all of the experiments. All except those whose numbers are above that of no. 78 were given daily feedings of iodine over a period of one to eleven days. The daily feeding consisted of a capsule containing two drops of tincture of iodine in starch. In the animals fed iodine the stimulation of the vago-sympathetic nerve was undertaken from two to ten days after the last feeding of iodine. Throughout the experiments the animals were *lightly* anesthetized with ether. Platinum wire electrodes were applied to opposite sides of the carefully isolated vago-sympathetic nerve and shielded from all surrounding tissues by glass. · In all of the experiments except those recorded in table 5, a tetanizing current from three to six times as strong as that sufficient to bring about a pupillary dilatation and apparent protrusion of the bulbus oculi was employed. The regulation of the strength of the current was made possible by a rheostat inserted in the circuit.   Again in all of the experiments except those to be found in table 5 the stimulating current throughout the period of stimulation was made for about 0.8 second at intervals of 1.6 seconds by means of a clock and ratchet device.   The strength of the current used and the rate at which the current was made in the experiments of table 5 are described below.

TABLE 1

Variations in the iodine content of neighboring specimens of the same lobe of the thyroid gland

ANIMAL NUMBER	NUMBER OF FEEDINGS OF IODINE	LOBE OF THYROID GLAND	WEIGHT OF SAMPLE OF DRIED GLAND ANALYZED	IODINE IN DRIED GLAND ANALYZED	DIFFERENCES IN THE CONCENTRATION OF IODINE IN NEIGHBORING SPECIMENS OF THE SAME LOBE
			gram	*per cent*	*per cent*
68	11	Left	0.1192	0.331	0.093
			0.1378	0.424	
		Right	0.1525	0.418	0.037
			0.1668	0.455	
74	6	Left	0.6931	0.366	0.005
			0.6420	0.371	0.034
			0.6846	0.405	0.039
		Right	0.4549	0.370	0.006
			0.4931	0.364	0.019
			0.5533	0.383	0.013
88	0	Left	0.4454	0.121	0.011
			0.5298	0.132	
		Right	0.7857	0.119	0.010
			0.4528	0.129	
105	0	Left	0.7241	0.158	0.018
			0.7138	0.176	
		Right	0.4489	0.163	0.015
			0.4516	0.148	

After a number of experiments had been performed it was found that there is considerable variation in the iodine content of adjoining pieces of the *same* gland.   In table 1, for example, are given a few analyses of neighboring specimens of the same gland.   The differences in neighboring portions of the glands are somewhat greater in iodine-fed animals.

TABLE 2

Quantitative determination of iodine in whole gland and in cells free from colloid material of thyroid glands of dogs whose isolated vago-sympathetic nerve had previously been stimulated for approximately three hours

ANIMAL NUMBER	LOBE STIMULATED	LENGTH OF PERIOD OF STIMULATION	WEIGHT OF WHOLE GLAND USED	IODINE IN WHOLE GLAND	WEIGHT OF CELL MASS	IODINE IN CELL MASS	RATIO OF PER CENT OF IODINE IN CELLS TO PER CENT OF IODINE IN WHOLE GLAND	APPARENT GAIN (+) OR LOSS (−) IN CONCENTRATION OF IODINE IN WHOLE GLAND STIMULATED LOBE
			gram	per cent	gram	per cent		per cent
42	Left	2 hrs. 15 min.	L 0.1792	0.531	0.2533	0.371	0.699	−0.029
			R 0.1693	0.560	0.1360	0.303	0.541	
45	Left	3 hrs. 0 min.	L 0.4602	0.148	0.5117	0.011	0.074	+0.046
			R 0.5336	0.102	0.4750	0.014	0.137	
46	Left	3 hrs. 30 min.	L 0.4982	0.098	0.3445	0.014	0.143	−0.006
			R 0.5252	0.104	0.2284	0.017	0.163	
54	Right	2 hrs. 30 min.	L 0.3964	0.177	0.2713	0.036	0.203	−0.030
			R 0.4241	0.147	0.3883	0.019	0.129	
55	Right	2 hrs. 40 min.	L 0.6880	0.232	0.2700	0.031	0.134	−0.046
			R 0.5509	0.186	0.2280	0.046	0.247	
56	Right	2 hrs. 10 min.	L 0.4536	0.121	0.3683	0.016	0.132	−0.012
			R 0.4795	0.109	0.3059	0.014	0.128	
62	Left	3 hrs. 5 min.	L 0.0836	0.494	0.0480	0.078	0.158	−0.029
			R 0.0558	0.523	0.0612	0.082	0.157	
64	Right	3 hrs. 50 min.	L 0.0892	0.110	0.1019	0.008	0.073	+0.017
			R 0.0692	0.127	0.1190	0.010	0.079	
65	Right	3 hrs. 0 min.	L 0.1179	0.436	0.1086	0.037	0.085	+0.030
			R 0.1419	0.466	0.1064	0.052	0.112	
66	Right	3 hrs. 0 min.	L 0.0709	0.485	0.0733	0.041	0.085	+0.008
			R 0.0593	0.493	0.0783	0.038	0.077	
67	Left	3 hrs. 0 min.	L 0.1373	0.464	0.1842	0.041	0.088	+0.013
			R 0.1237	0.451	0.1613	0.040	0.089	
68	Left	2 hrs. 45 min.	L 0.2570	0.377	0.1703	0.031	0.082	−0.060
			R 0.3193	0.437	0.1379	0.028	0.064	
69	Left	3 hrs. 0 min.	L 0.5360	0.605	0.2627	0.083	0.137	+0.036
			R 0.3957	0.569	0.1476	0.092	0.162	
71	Right	3 hrs. 0 min.	L 0.3651	0.346	0.1718	0.026	0.075	−0.041
			R 0.2990	0.305	0.1310	0.029	0.095	

Hence it was thought desirable to analyze not single blocks or samples of dried powdered mixtures of whole gland but to analyze the *whole* gland in each case. In experiments in which this last mentioned technique was employed the whole gland was carefully cleaned of connective tissue and blood vessels, and thoroughly dried first over an electric hot plate and then in an electric oven. After two to three hours' drying in the electric oven the gland was broken into several pieces whose weight was about 0.5 gram each and whose number therefore depended on the size of the gland. The iodine determinations were made according to the method earlier described by Kendall (13). By analysis of powdered thyroid of known iodine content the accuracy of the method (to about 0.008 mgm. of iodine) and the purity of the reagents used were frequently examined and found to be satisfactory.

The strength of the current employed in all of the experiments recorded in table 2 was three times that sufficent to cause dilatation of the pupil and apparent protrusion of the bulbus oculi. Both the stimulated and the non-stimulated vago-sympathetic nerves were cut in two places: at a point in the neighborhood of the eighteenth tracheal ring and also at a level a little above that of the hyoid bone. The reason for cutting the non-stimulated nerve in such a manner was to eliminate the possible effect of tonic secretory impulses on the non-stimulated lobe. The vago-sympathetic nerve was stimulated a little above the point at which it was cut low in the neck. In the above experiments the ratio of the percentage of iodine in cells to the percentage of iodine in whole gland was determined by a method previously described (14). From the data given in table 2 it may be seen that stimulation of the vago-sympathetic nerve under the conditions described is without appreciable effect on either the ratio value or the concentration of iodine in the whole gland.

When it was found that there was no consistent diminution in the concentration of iodine in the stimulated lobe only the stimulated vago-sympathetic nerve was sectioned in the manner described above. In experiments 76, 79 and 80 a strength of current six times that necessary to cause ocular changes characteristic of sympathetic stimulation was used; in all of the other experiments to be found in table 3 the current was of the same strength as that used in the experiments recorded in table 2. From the standpoint of the iodine concentration in whole gland the results given in tables 2 and 3 are very similar. Stimulation apparently has no effect on the concentration of iodine in the stimulated lobe.

By stimulating the intact vago-sympa'thetic nerve Rahe et al. (2) declare that they were able to produce the most marked diminution in the iodine content of a given lobe of the thyroid gland. In the three experiments of table 4 of my series, the intact vago-sympathetic nerve was

TABLE 3

The concentration of iodine in the lobes of the thyroid gland of the dog after the stimulation of the isolated vago-sympathetic nerve on one side for a period of approximately three hours

ANIMAL NUMBER	LOBE STIMU-LATED	LENGTH OF PERIOD OF STIMULATION	WEIGHT OF DRIED WHOLE GLAND		IODINE IN DRIED WHOLE GLAND	APPARENT GAIN (+) OR LOSS (−) IN CONCENTRA-TION OF IODINE IN STIMULATED LOBE
				gram	*per cent*	*per cent*
72	Right	2 hrs. 50 min.	L	1.2883	0.186	+0.003
			R	1.4258	0.189	
74	Right	3 hrs. 10 min.	L	2.0197	0.381	−0.008
			R	1.5013	0.373	
75*	Left	3 hrs. 10 min.	L	1.4975	0.334	−0.026
			R	1.8796	0.360	
76	Right	3 hrs. 10 min.	L	0.4653	0.770	+0.051
			R	0.3509	0.821	
77	Right	3 hrs. 30 min.	L	0.7055	0.471	+0.037
			R	0.7914	0.508	
79	Right	3 hrs. 15 min.	L	0.2421	0.474	+0.024
			R	0.1995	0.498	
80	Left	4 hrs. 0 min.	L	0.6342	0.465	−0.012
			R	0.4850	0.477	
82	Left	3 hrs. 10 min.	L	0.8483	0.203	−0.003
			R	0.7516	0.206	
83	Left	3 hrs. 15 min.	L	1.2435	0.105	+0.002
			R	0.9163	0.103	
84	Left	3 hrs. 10 min.	L · 0.7582		0.129	−0.016
			R	0.6022	0.145	

* Entire right lobe not analyzed.

stimulated in the neck at about the level of the fifteenth tracheal ring. The current was of the same strength as that employed in the experiments of table 2; each time the current was made there ensued a respiratory arrest and the ocular changes typical of sympathetic stimulation. Again there was no consistent change in the concentration of iodine in the stimulated lobe.

TABLE 4

The concentration of iodine in the lobes of the thyroid gland of the dog after the stimulation of the intact vago-sympathetic nerve on one side for a period of approximately three hours

ANIMAL NUMBER	LOBE STIMU- LATED	LENGTH OF PERIOD OF STIMULATION	WEIGHT OF DRIED WHOLE GLAND	IODINE IN DRIED WHOLE GLAND	APPARENT GAIN (+) OR LOSS (−) IN CONCENTRA- TION OF IODINE IN STIMULATED LOBE
			gram	*per cent*	*per cent*
78	Left	3 hrs. 40 min.	L 0.4628	0.528	+0.027
			R 0.3706	0.501	
86	Right	3 hrs. 20 min.	L 0.7039	0.138	−0.013
			R 0.6919	0.125	
88	Left	3 hrs. 15 min.	L 0:9752	0.127	+0.004
			R 1.2385	0.123	

Effect of vasomotor activity on the concentration of iodine in the thyroid gland. It will be recalled that Watts (3) concluded from his experiments that vascular changes will account for the diminution in the concentration of iodine which he brought about by stimulation of the cervical sympathetic nerve. The experiments of table 5 were undertaken to find out whether or not a slightly different type of stimulus sent into the nerve at an interval more nearly like that employed by Watts had an effect comparable to that found in the experiments previously performed. It was also thought desirable to determine whether or not the characteristic vascular changes were present throughout the experiments.

The technique except for certain features of the stimulation was the same as that used in all of the preceding experiments. Usually the stimulated vago-sympathetic nerve was ligated and cut at about the level of the eighteenth to twentieth tracheal ring; near the ganglion nodosum only the vagus nerve was cut. In all of the experiments the sympathetic chain was intact above the eighteenth tracheal ring. The

TABLE 5

The concentration of iodine in the lobes of the thryoid gland of the dog after the stimulation of the vago-sympathetic nerve, the sympathetic portion of which was left intact above the point of stimulation

ANIMAL NUMBER	LOBE STIMULATED	LENGTH OF PERIOD OF STIMULATION	PRESENCE (+) OR ABSENCE (−) OF VASOMOTOR ACTIVITY AT END OF EXPERIMENT	WEIGHT OF DRIED WHOLE GLAND		IODINE IN DRIED WHOLE GLAND	APPARENT GAIN (+) OR LOSS (−) IN CONCENTRATION OF IODINE IN STIMULATED LOBE
					grams	per cent	per cent
101	Right	3 hrs. 10 min.	−	L	1.8213	0.177	−0.052
				R	1.5890	0.125	
102	Left	3 hrs. 0 min.	+	L	0.1397	0.141	+0.021
				R	0.1727	0.120	
103	Right	3 hrs. 0 min.	+	L	0.6335	0.277	−0.003
				R	0.5749	0.274	
104	Right	3 hrs. 0 min.	−	L	0.1891	0.190	−0.026
				R	0.1844	0.164	
105	Right	3 hrs. 25 min.	+	L	1.4379	0.167	−0.011
				R	0.9005	0.156	
106	Left	3 hrs. 25 min.	+	L	0.6161	0.043	+0.002
				R	0.6624	0.041	
107	Right	3 hrs. 0 min.	−	L	3.1065	0.103	+0.006
				R	2.9642	0.109	
108	Left	3 hrs. 35 min.	+	L	0.3462	0.021	−0.006
				R	0.2754	0.027	
109	Right	3 hrs. 25 min.	+	L	1.0407	0.014	+0.002
				R	1.0116	0.016	
111	Left	3 hrs. 50 min.	+	L	0.2875	0.076	+0.001
				R	0.2638	0.075	
113	Right	3 hrs. 30 min.	−	L	1.5649	0.034	−0.001
				R	2.6585	0.033	
114	Right	3 hrs. 30 min.	+	L	0.3210	0.143	−0.011
				R	0.3690	0.132	
115	Left	3 hrs. 45 min.	+	L	1.0381	0.088	−0.002
				R	0.8196	0.090	
116	Left	3 hrs. 30 min.	−	L	0.6650	0.016	−0.001
				R	0.6753	0.017	
117	Left	4 hrs. 20 min.	+	L	1.1050	0.140	+0.006
				R	1.1701	0.134	
118	Right	5 hrs. 0 min.	−	L	0.6568	0.456	−0.018
				R	0.5036	0.438	
119	Right	3 hrs. 10 min.		L	0.6866	0.168	+0.005
				R	0.5695	0.173	

vago-sympathetic nerve was stimulated just above the point at which it was cut near the eighteenth tracheal ring. It was easily possible to vary the strength of the electrical stimulus so that a current of such a strength was employed as just to bring about the ocular changes characteristically associated with stimulation of the cervical sympathetic nerve. Mendenhall (15) has emphasized the markedly toxic effects of ether on the sympathetic nervous system. In a preparation like that last described the sensitivity of the cervical sympathetic nerve to the depressant action of ether could easily be demonstrated. A tetanizing current of rather low frequency from a Stoelting inductorium was made for 5.5 seconds at intervals of 11.8 seconds throughout the period of stimulation. At the end of most of these experiments a vein of the stimulated lobe, in most cases without the ligation of its companion veins, was cannulated and the effect on blood flow of stimulation of the gland under the *same* conditions as those employed in the experiment was observed. In ten out of sixteen experiments no difficulty was encountered in demonstrating a vasoconstriction in the gland on stimulating the vago-sympathetic nerve with an electric current of the same strength and delivered at the same rate as that used in the previous stimulation period. The threshold of excitability for the vasoconstrictor nerves of the thyroid gland appears to be considerably lower than that for the submaxillary gland as reported by Gruber (16). The relatively low threshold of excitability of the vasoconstrictor nerves of the thyroid to epinephrin stimulation has been observed by Gunning (17).

From table 5 it may be seen that stimulation of the vago-sympathetic nerve with the sympathetic trunk intact above the eighteenth tracheal ring has no appreciable effect on the concentration of iodine in the stimulated lobe. In a number of experiments it was unequivocally demonstrated that vasoconstrictor fibers to the gland were being stimulated; yet such stimulation did not alter detectably the iodine content of the lobe subjected to stimulation.

Effect of stimulation of the vago-sympathetic nerve on the concentration of water in the thyroid gland. In table 6 may be found data relative to the effect of stimulation of the vago-sympathetic nerve on the concentration of water in the thyroid gland. I was unable to find that stimulation had any effect on the concentration of water in the stimulated lobe even in experiments in which vasomotor changes were definitely proved to result from stimulation.

TABLE

The concentration of water in the lobes of the thyroid gland of the dog afte lation of the vago-sympathetic nerve for a period of from three to hours

ANIMAL NUMBER	LOBE STIMULATED	LENGTH OF PERIOD OF STIMULATION	CONCENTRATION OF WATER	APPARENT GAIN (+) OR LOSS (−) IN CON- WATER IN STIMU- LATED LOBE
			per cent	
77	Right	3 hrs. 30 min.	L 69.25 R 69.64	+0.39
79	Right	3 hrs. 15 min.	L 73.40 R 74.26	
82	Left	3 hrs. 10 min.	L 74.06 R 74.66	−0.60
83	Left	3 hrs. 15 min.	L 77.57 R 78.02	−0.45
84	Left	3 hrs. 10 min.	L 74.38 R 74.24	+0.14
88	Left	3 hrs. 15 min.	L 73.65 R 72.81	+0.84
108*	Left	3 hrs. 35 min.	L 80.14 R 80.71	−0.57
109*	Right	3 hrs. 25 min.	L 78.54 R 80.64	+2.10
110*	Right	3 hrs. 45 min.	L 82.69 R 82.13	−0.56
111*	Left	3 hrs. 50 min.	L 78.03 R 77.33	+0.70
113	Right	3 hrs. 30 min.	L 75.98 R 73.73	−2.25
115*	Left	3 hrs. 45 min.	L 75.76 R 75.63	+0.13
116	Left	3 hrs. 30 min.	L 79.92 R 78.34	+1.58
117*	Left	4 hrs. 20 min.	L 76.53 R 75.65	+0.88
119	Right	3 hrs. 10 min.	L 75.90 R 76.81	+0.91

* Vasomotor effect of the stimulation demonstrated at the end of the experiment.

DISCUSSION

If we consider the data described above we find that following the stimulation of a given lobe of the thyroid gland there was no consistent change in the concentration of iodine in that lobe compared with the

TABLE 7

A comparison of the effect of the different methods of stimulation of the vago-sympathetic nerve on the concentration of iodine in the thyroid gland of the dog

SERIES	NON-SIGNIFICANT	APPARENT LOSS IN CONCENTRATION OF IODINE			APPARENT GAIN IN CONCENTRATION OF IODINE			REMARKS
		Number	Maximum	Average	Number	Maximum	Average	
			per cent	per cent		per cent	per cent	
I	2	7	0.060	0.035	5	0.046	0.028	Stimulated vago-sympathetic nerve cut above hyoid bone and low in the neck. Non-stimulated vago-sympathetic nerve cut similarly in most cases
II	4	3	0.026	0.018	3	0.051	0.037	Only stimulated vago-sympathetic nerve cut as in series I
III	1	1	0.013	0.013	1	0.027	0.027	Intact vago-sympathetic nerve stimulated at about level of fifteenth tracheal ring
IV	11	5	0.052	0.024	1	0.021	0.021	Vago-sympathetic nerve stimulated at about level of eighteenth tracheal ring. Sympathetic nerve entirely intact above point of stimulation
V	7	2	0.011	0.011	1	0.021	0.021	Animals of series IV in which vasomotor effect of stimulation could be shown plainly at end of experiment

nonstimulated lobe. Following stimulation there was an apparent diminution in the concentration of iodine in 24 or 54.5 per cent of the stimulated glands, and an apparent gain in 20 or 45.5 per cent of glands stimulated similarly. Watts (3) reported that the average difference in the percentage of iodine in the lobes of dried thyroid gland of the dog is 0.015 per cent in this vicinity. If we consider as non-significant all differences in the percentage of iodine in dried thyroid of 0.010 per cent or less we see that stimulation, while causing an apparent diminution in the concentration of iodine in 16 or 36.4 per cent of the glands, and an apparent increase in the concentration of iodine in 10 or 22.7 per cent of the glands, had no effect on the iodine concentration of 18 or 40.9 per cent of the stimulated glands. The greater part of the experimental data presented may be briefly summarized in table 7.

The differences which I found above appear to be due to normal variations in the concentration of iodine in the two lobes of the dog's thyroid gland. The average percentage difference in the concentration of iodine in the two lobes of the dog's thyroid gland depends on a number of factors such as the type of gland, the time of year and the feeding of iodine. Hence I am forced to conclude that if stimulation of the vago-sympathetic nerve in the dog has any effect on the concentration of iodine in the thyroid gland, that effect is considerably less than the normal variation in the iodine content of the stimulated and control lobes. The presence of normal variations relatively so much greater than the variations which may follow stimulation renders valueless the application of more refined methods of iodine determination in the study of the effect of stimulation of the cervical sympathetic nerve on the concentration of iodine in the thyroid gland.

SUMMARY

1. Periodic stimulation of the isolated vago-sympathetic nerve by an induced current of a moderate to strong intensity over a period of from three to three and a half hours does not appreciably alter the distribution ratio of iodine between cells and colloid.

2. The findings of Rahe, Rogers, Fawcett and Beebe (2) and of Watts (3) that stimulation of the cervical sympathetic nerve for a comparable period of time reduces the concentration of iodine in the stimulated lobe were not confirmed. Watts' assertion that such stimulation reduces the water content of the stimulated lobe was not confirmed.

3. Conclusions as to the direct secretory control which the cervical sympathetic nerves exercise on the thyroid gland are based in no small measure on the alleged effect of stimulation of the cervical sympathetic nerve on the iodine content of the gland. Conclusions having such a basis apparently are untenable.

It is a pleasure to acknowledge the many suggestions and constant aid given me by Dr. A. L. Tatum during the progress of this work.

• BIBLIOGRAPHY

(1) WIENER: Arch. exper. Path. u. Pharm., 1909, lxi, 297.

(2) RAHE, ROGERS, FAWCETT AND BEEBE: This Journal, 1914, xxxiv, 72.

(3) WATTS: This Journal, 1915, xxxviii, 356.

(4) CANNON, BINGER AND FITZ: This Journal, 1915, xxxvi, 363.
CANNON AND FITZ: This Journal, 1916, xl, 126.

(5) CANNON AND CATTELL: This Journal, 1916, xli, 58, 74.

(6) CANNON AND SMITH: Endocrin., 1920, iv, 386.

(7) LEVY: This Journal, 1916, xli, 492.

(8) BURGET: This Journal, 1917, xliv, 492.

(9) MARINE, ROGOFF AND STEWART: This Journal, 1918, xlv, 268.

(10) TROELL: Arch. Int. Med., 1916, xvii, 382.

(11) MILLS: This Journal, 1919, l, 174.

(12) ROGOFF: Journ. Pharm. Exper. Therap., 1918, xii, 193.

(13) KENDALL: Journ. Biol. Chem., 1914, xix, 251.

(14) TATUM: Journ. Biol. Chem., 1920, xlii, 47.

(15) MENDENHALL: This Journal, 1914, xxvi, 57.

(16) GRUBER: This Journal, 1915, xxxvii, 259.

(17) GUNNING: This Journal, 1917, xliv, 215.

Reprinted from THE JOURNAL OF BIOLOGICAL CHEMISTRY
Vol. LIV, No. 1, September, 1922

A STUDY OF THE DISTRIBUTION OF IODINE BETWEEN CELLS AND COLLOID IN THE THYROID GLAND.

IV. THE DISTRIBUTION OF IODINE IN THE HYPERPLASTIC THY-ROID GLAND OF THE DOG AFTER THE INTRAVENOUS INJECTION OF IODINE COMPOUNDS.

BY HARRY BENJAMIN VAN DYKE.

(From the Laboratories of Pharmacology, University of Chicago, Chicago.)

(Received for publication, July 20, 1922.)

Some years ago Marine and Feiss (1) and Marine and Rogoff (2, 3) first performed experiments which leave little doubt as to the ability of the dog's thyroid gland, especially when hyperplastic, to bind iodine almost instantaneously. Marine and Feiss (1) carefully perfused the surviving thyroid gland with fluid containing iodine as KI. They found that after 1 hour a considerable amount of iodine was taken up only by a *surviving* gland. Any evidence of death of the perfused organ was accompanied by a loss of some of the gland's stored iodine rather than by an absorption of iodine from the circulating medium. However, even a surviving gland, rich in iodine, lost some iodine to a perfusing fluid free of the element. In surviving glands they discovered that the amounts of iodine absorbed were relatively independent of the amounts of iodine in the perfusing fluids. They also pointed out that a similarly rapid absorption of iodine by the intact gland follows the intravenous administration of a solution of KI. From the results of perfusions of spleen and kidney under similar conditions they concluded that these organs were not capable of taking up a significant amount of iodine.

Marine and Rogoff (2) on the basis of experiments in which they injected a solution of KI intravenously came to the conclusion that the absorption of iodine by the gland is almost as great 1 hour after the injection as it is 30 hours after the injection. They again found that no significant amount of iodine was taken up by the spleen and liver. The amount of iodine absorbed by the

11

thyroid gland appeared to be directly proportional to the degree of hyperplasia exhibited by the gland. In a second communication (3) they discussed the time of appearance of the changes in the histology and physiological activity of the gland following the intravenous administration of KI solution.

Having found (4) that there was relatively little difference in the ratio of the percentage of iodine in cells to the percentage of iodine in whole gland in dog thyroid glands exhibiting great variations in histological appearance and iodine content, I undertook the present study to determine what effect acute iodization of hyperplastic thyroid glands has on the ratio value.

Methods.

As in the work of Marine and Feiss, and Marine and Rogoff, dogs with thyroid glands usually definitely hyperplastic were used in all of the experiments. Light ether narcosis was always employed. All solutions of KI and thyroid colloid were injected into the femoral vein. Every effort was made to section the glands as rapidly as possible after their removal from the animal. In the making of the determinations of the ratio of the percentage of iodine in dried cells to that in dried whole gland the method first described by Tatum (5) was somewhat modified. To lessen autolysis as much as possible the glands on removal were plunged into Ringer's solution cooled to 1–4°C. The glands were then carefully cut into blocks of a size appropriate for the floor of the freezing microtome and frozen sections were made of a considerable amount of tissue. The blocks of tissue cut for the microtome's floor as well as the cut sections were suspended in Ringer's solution cooled to 1–4°C. As in the preceding work of this series the frozen sections were cut sufficiently thin so that practically all of the colloid or intercellular fluid was dissolved out of the acini as soon as the sections were suspended in cooled Ringer's solution. By centrifugalization the cells were separated from the colloid dissolved in the Ringer's solution and carefully dried over an electric hotplate and then in an electric drying oven. Control pieces of uncut whole gland were also dried in the same manner. Throughout this paper colloid iodine solutions refer to cell-free solutions of thyroid intercellular fluids obtained by the method just described and dissolved in Ringer's solution. In all of the experiments the quantitative determinations of iodine were made by the method of Kendall (6).

From the data of Table I it can be seen that large amounts of iodine as KI were taken up by hyperplastic thyroid glands despite big variations in the dose of KI (50 to 150 mg.) and duration of time between injection and removal of gland (90 seconds to 22 hours and 25 minutes). The cooling of the blocks of thyroid tissue during the process of cutting had little effect. Either little of the injected iodine was held by the cells, or it diffused rapidly from the cells in the process of cutting; for the iodine content of the cells and hence the ratio of the precentage of iodine in cells to the percentage of iodine in whole gland is considerably less than that found in resting glands. The average ratio value of 0.15 for the dog (4) was approached only in Experiment 11 in which nearly 24 hours elapsed between the injection of KI and the removal of the experimental lobe. In this last case there are alternative means of explanation: more iodine may have been bound by the cells by removal from the thyroid's intercellular spaces, or the iodine in the cells was in a less diffusible form than in the other experiments of the series. These changes in the distribution of iodine after about 24 hours can be definitely correlated with Marine and Rogoff's (3) discovery that 20 hours after the injection of iodine as KI, markedly hyperplastic glands exhibit more stainable colloid with some increase in the size of the follicular spaces and shrinkage in the height of the columnar epithelium.

In the experiments the data of which are given in Table II, solutions of colloid iodine from normal dog thyroid glands were injected intravenously to determine to what extent iodine in organic combination is bound by hyperplastic thyroid glands.

The iodine of thyroid colloid of normal animals is practically not at all taken up by hyperplastic glands after 42 to 86 minutes. A comparable amount of iodine in the form of KI is rapidly taken up.

Blum and Grützner (7) found some hours after the injection into the circulatory system of fluid pressed from thyroid glands that the iodine-containing thyroid protein is split largely by hepatic action into simpler products with the formation even of iodides. In Experiments 12 and 13 (42 and 86 minutes after intravenous injection) there is little evidence that much iodide-

TABLE I.

Ratio of the Percentage of Iodine in Cells to the Percentage of Iodine in Whole Glands in Hyperplastic Thyroid Glands of Dogs Receiving an Intravenous Injection of KI Solution for Varying Lengths of Time before the Removal of the Glands.

Animal No.	Amount of KI injected.	Time elapsing between injection and removal of gland.	Lobe.	Weight of whole gland used.	Iodine in whole gland.	Weight of cell mass used.	Iodine in cell mass.	Ratio of per cent of iodine in cells to per cent of iodine in whole gland.
	mg.	sec.		gm.	per cent	gm.	per cent	
1	50	90	Control.	0.6786	0.021	0.5242	0	
			Iodized.	0.5553	0.042	0.7845	0.002	0.048
2	50	150	Control.	0.6548	0.010	0.6373	0	
			Iodized.	0.5695	0.072	0.8437	0.003	0.042
		min.						
3*	135	57	Control.	0.6797	0	0.3585	0	
			Iodized.	0.4808	0.086	0.5449	0	
4	120	60	Control.	0.4512	0.042	0.3775	0	
			Iodized.	0.7760	0.079	0.4702	0	
5	150	60	Control.	0.6853	0	0.6105	0	
			Iodized.	0.6302	0.078	0.7473	0	
6	50	60	Control.	0.8779	0.002	0.5694	0	
			Iodized.	0.8314	0.059	0.8021	0.004	0.068
7	50	60	Control.	0.8130	0			
			Iodized.	0.6521	0.085	0.6994	0.002	0.025
				0.7311	0.070			
8	50	60	Control.	0.4082	0.007			
				0.6490	0.004			
		•	Iodized.	0.4610	0.081	0.8222	0.004	0.045
				0.2455	0.098			
9	150	64	Control.	0.5150	0.002	0.7066	0	
			Iodized.	0.5957	0.085	0.8282	0.003	0.035
10*	120	70	Control.	0.3458	0.010	0.2459	0	
			Iodized.	0.2502	0.091	0.2627	0	
		hrs. min.						
11	50	22 25	Control.	0.4650	0.017	0.5357	0	
				0.5715	0.020			
			Iodized.	0.7191	0.213	0.5422	0.022	0.103

* Glands and cells not suspended in ice-cooled Ringer's solution.

14

iodine has been split from colloid iodine; for the hyperplastic thyroid glands present exhibit no significant change in iodine content and yet are able quickly to bind any iodine available as iodide.

In the experiments reported in Table III colloid iodine solution of animals with hyperplastic glands, each of which had received an

TABLE II.

Ratio of the Percentage of Iodine in Cells to the Percentage of Iodine in Whole Gland in Hyperplastic Thyroid Glands of Dogs after the Intravenous Injection of Thyroid Colloid Iodine from Normal Dog Thyroid Glands. The Iodine Content of a Hyperplastic Gland of a Dog after the Injection of a Comparable Amount of Inorganic Iodine is also Recorded.

Animal No.	Amount of solution injected.	Form of iodine.	Amount of iodine.	Time elapsing between completion of injection and removal of iodized lobe.	Lobe.	Weight of whole gland used.	Iodine in whole gland.	Weight of cell mass used.	Iodine in cell mass.	Ratio of per cent of iodine in cells to per cent of iodine in whole gland.
	cc.		mg.	min.		gm.	per cent	gm.	per cent	
12	200.0	Colloid iodine.	1.690	42	Control.	0.6565	0.002	0.5478	0.001	
					Iodized.	0.6248	0.002	0.4819	0	
13	259.6	Colloid iodine.	2.829	86	Control.	0.7971	0	0.7698	0	
						0.8557	0			
					Iodized.	0.4742	0.005	0.9208	0	
						0.4245	0			
14	241.5	KI	2.087	46	Control.	0.7939	0.003			
						0.6024	0.004			
					Iodized.	0.4513	0.017			
						0.5907	0.019			

intravenous injection of 50 mg. of KI 60 minutes before the removal of the thyroid lobe for section, was injected intravenously as soon as possible into other animals with hyperplastic glands. The sequence was as follows: as a control, part of one lobe was removed; colloid iodine dissolved in Ringer's solution was then injected and all but half of one lobe resected; a comparable amount of iodine as KI dissolved in a similar amount of Ringer's solution was then

TABLE III.

Ratio of the Percentage of Iodine in Cells to the Percentage of Iodine in Whole Gland in Hyperplastic Thyroid Glands of Dogs after the Intravenous Injection of Thyroid Colloid Iodine from Hyperplastic Thyroid Glands Iodized but a Short Time before Being Sectioned.

Animal No.	Form and amount of iodine injected.	Time elapsing between injection and removal of lobe sectioned or analyzed.	Lobe.	Weight of whole gland used.	Iodine in whole gland.	Weight of cell mass used.	Iodine in cell mass.	Ratio of per cent of iodine in cells to per cent of iodine in whole gland.
		min.		gm.	per cent	gm.	per cent	
15	0.820 mg. of colloid iodine of Animal 7* dissolved in 233 cc. of Ringer's solution.	60	Control. Colloid iodine.	0.6022 0.6986 0.7186	Trace. 0.009 0.009	1.1376	0	
	0.836 mg. of KI dissolved in 253.5 cc. of Ringer's solution.	25	KI	0.8328 0.6935	0.017 0.022			
16	0.903 mg. of colloid iodine of Animal 8* dissolved in 318 cc. of Ringer's solution.	40	Control. Colloid. iodine.	0.4700 0.4015 0.3695	0 0.006 0.007	1.1110	0	
	0.827 mg. of KI dissolved in 318 cc. of Ringer's solution.	6	KI	0.4581 0.4437	0.009 0.010			

* See Table I.

injected and the remaining half of one lobe taken out. Much more of the iodine from the colloid iodine of these animals was taken up than from the colloid iodine of normal animals (Table II).

Even here considerable iodine as KI was still bound by the glands which had taken up some colloid iodine. These results suggest that the colloid iodine, bound but incompletely synthesized into active principle, is in a more diffusible form in the acutely iodized gland than in the more normal resting gland. Moreover, the incompletely synthesized active principle may be more readily split by hepatic action (7) into simpler products (*e.g.*, iodides) which are then bound by the hyperplastic thyroid gland.

Marine and Rogoff (3) declare that even 30 hours after intravenous administration of a solution of KI very little of the thyroid's active principle, as measured by the effects of thyroid on tadpole growth and metamorphosis, has been elaborated. The less diffusible and less readily split normal colloid iodine compound (Table II) probably represents the more fully elaborated active principle.

SUMMARY.

The findings of Marine and Feiss and Marine and Rogoff that the hyperplastic thyroid gland of the dog rapidly binds iodine intravenously introduced as a solution of KI were confirmed.

By a method already described (5) the ratio value of iodine in cells to iodine in whole gland was determined and found to be very low after the intravenous injection of KI solution into dogs with hyperplastic glands when those glands were removed 1.5 to 60 minutes after the injection. The ratio value more nearly approached the normal if the interval elapsing between injection and removal of gland was made about 24 hours instead of 1 hour or less as in most of the experiments. This finding is in keeping with the histological changes described by Marine and Rogoff in such glands 20 hours after the intravenous injection of KI solution.

When iodine as colloid iodine solution of normal animals was administered intravenously practically none of the colloid iodine was taken up by hyperplastic glands during the periods of time used in these experiments; yet from an injection of a comparable amount of iodine in the form of KI the ready binding of iodine by similarly hyperplastic glands was proved. Colloid iodine of hyperplastic glands removed 1 hour after the intravenous injection of KI solution was taken up to some extent by hyperplastic glands; but these last named glands bound additional iodine as KI introduced after the colloid iodine injection. The incompletely synthesized active principle is probably more diffusible and more readily split into simpler products than active principle fully synthesized.

BIBLIOGRAPHY.

1. Marine, D., and Feiss, H. O., J. *Pharmacol. and Exp.* Therap., 1915, vii, 557.
2. Marine, D., and Rogoff, J. M., J. *Pharmacol. and Exp.* Therap., 1916, viii, 439.
3. Marine, D., and Rogoff, J. M., J. *Pharmacol. and Exp.* Therap., 1916–17, ix, 1.
4. van Dyke, H. B., J. *Biol.* Chem., 1920–21, xlv, 325.
5. Tatum, A. L., J. *Biol.* Chem., 1920, xlii, 47.
6. Kendall, E. C., J. *Biol.* Chem., 1914, xix, 251.
7. Blum, F., and Grützner, R., Z. *physiol.* Chem., 1920, cx, 277.

**THIS BOOK IS DUE ON THE LAST DATE
STAMPED BELOW**

AN INITIAL FINE OF 25 CENTS

WILL BE ASSESSED FOR FAILURE TO RETURN
THIS BOOK ON THE DATE DUE. THE PENALTY
WILL INCREASE TO 50 CENTS ON THE FOURTH
DAY AND TO $1.00 ON THE SEVENTH DAY
OVERDUE.